SECTION ONE

HUMAN PHYSIOLOGY

1   THE ORGANISATION OF THE HUMAN BODY

The basic building blocks of the human body are called cells. Each cell is a living structure too small to be seen with the naked eye. With the aid of an optical microscope, found in a college or school laboratory, we can see that a human cell has an outer cell membrane, a jelly-like cytoplasm enclosed within the outer membrane and a small, central round nucleus.

Cell membrane

This regulates the entry and exit of chemicals in and out of the cell. It also holds the cell together. The cell membrane also adheres to other cells. It consists mainly of a double layer of molecules called phospholipid molecules, with protein molecules spanning it. It is also called the plasma membrane.

Cytoplasm

This appears as a jelly-like mass inside the cell. However, with a powerful electron microscope we can see small membrane-bound structures within it called organelles.

Organelles

- There are oval-shaped structures called mitochondria which are the site of production of high energy ATP molecules. They are called the power houses of the cell.
- Throughout the cytoplasm are interconnected membranes called the endoplasmic reticulum, which form a network of canals. These distribute chemicals around the cell. Rough endoplasmic reticulum has granular ribosomes on it. These are the sites where proteins are manufactured by the cell.
- The smooth endoplasmic reticulum is where lipids are made.
- The Golgi body can also be seen. This is a series of hollow, flattened discs with vesicles emerging from them. Its role is to package secretions for export from the cell, e.g. mucus.
- The lysosomes, which are spherical bodies that are the waste disposal units of the cell. They contain powerful hydrolytic enzymes which can destroy unwanted structures in the cell.
- The centrioles, two small cylinders located near the nucleus, which make spindle fibres in the cell during cell division, to move the chromosomes.

The nucleus

This is the container for the genetic instructions for the cell. These consist of DNA molecules which hold the genetic code. Just before cell division, the DNA condenses into 23 pairs of thread-like chromosomes. The nucleus also contains the nucleolus which makes ribosomes.

Introduction

This book is designed to give an introduction to human biology for students at Colleges of Further Education studying the Health pathways on Access to Higher Education Diploma courses. My students in the past have needed an introductory book in biology to prepare them for the ACCESS biology course. It is based on the units set by OCN for their biology module. A simplified approach has been adopted, making it suitable for all human biology students. I hope it may also be of interest to the non-academic reader who wishes to learn more about the human body and its functioning.

John White, BSc (Hons) Biological Sciences, University of Aston in Birmingham, PGCE, Cert. Health Ed.

Former Lecturer in Biology, Wiltshire College, Salisbury

Acknowledgements

I would like to thank Robert George of SSER Ltd for his kind support, Ross Poole for help with electronic editing and also Peta Reid of Sydney University.

Diagrams

The publishers and author wish to thank SSER Ltd for permission to use their diagrams.

Title picture of DNA and figures 1.1, 2.1, 2.2, 3.1, 3.2, 3.4, 3.5, 3.6, 4.1, 4.2, 9.1, 11.1, 11.2, 13.6– SSER Ltd.- www.sserltd.co.uk, Telephone 01404 811 667.

Also OpenStax College for figure 3.3- OpenStax College 2013 Cardiac Cycle, Connexions, June 19, 2013 http://cnx.org/content/m46661/1.3/.

CONTENTS

SECTION ONE: HUMAN PHYSIOLOGY

Chapter One: The Organisation of the Human Body — Pages 6-7

Chapter Two: The Respiratory system — Pages 8-10

Chapter Three: The Circulatory system — Pages 11-15

Chapter Four: The Digestive system — Pages 16-19

SECTION TWO: DEFENCE AGAINST DISEASE

Chapter Five: Diseases — Pages 22-23

Chapter Six: Immunology– The Non-specific response — Page 24

Chapter Seven: Immunology-The Specific response — Pages 25-26

SECTION THREE: HOMEOSTASIS

Chapter Eight: The Principles of Homeostasis — Page 28

Chapter Nine: Temperature Regulation — Pages 29-31

Chapter Ten: Glucose Regulation — Pages 32-33

SECTION FOUR: GENES AND GENE TECHNOLOGY

Chapter Eleven: DNA and Genes — Pages 36-38

Chapter Twelve: Cell Division — Pages 39-41

Chapter Thirteen: Genetics — Pages 42-49

Chapter Fourteen: Gene Technology — Pages 50-51

Figure 1.1 A generalised human cell

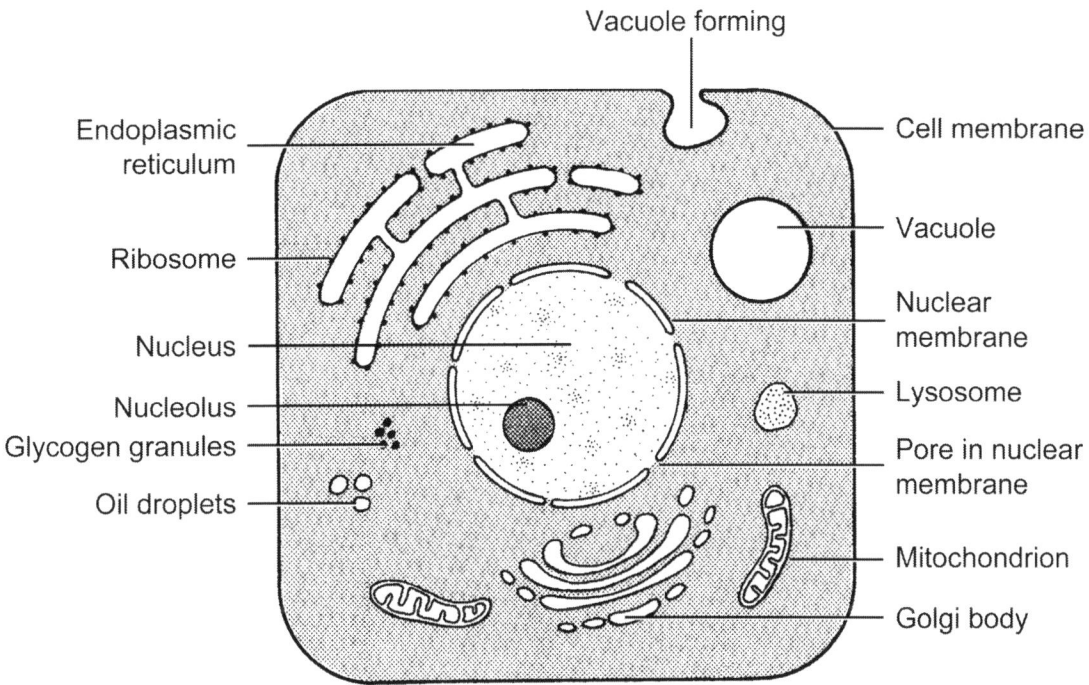

Cells, tissues, organs and organ systems

The cells, as we have seen, are the basic building blocks of the human body. They can be specialised into different types of cells to carry out different functions in the body, e.g. muscle cells contain filaments that can move past each other causing the cells to contract. Nerve cells have long fibres extending from them, which carry electrical nerve impulses and sperm cells have a whip-like flagellum or tail, which can propel the sperm cell.

Groups of similar cells that are specialised to work together to perform a particular function form **tissues**, e.g. muscle tissue and bone tissue.

Groups of tissues may form **organs** with a particular function, e.g. the heart is made of muscle tissue, connective tissue, epithelial tissue and nervous tissue.

Organs work together in groups called **organ systems**, e.g. the heart, blood vessels and blood form the circulatory system; the oesophagus, stomach, pancreas, liver, gall bladder and intestines form the digestive system.

Organ systems combine together to form the **human body**, e.g. the circulatory system, the nervous system, the musculo-skeletal system, the digestive system, the reproductive system, the respiratory system, the excretory system and the endocrine system.

2   THE RESPIRATORY SYSTEM

The breathing system includes the trachea, lungs, bronchi, bronchioles, alveoli, diaphragm, ribs and intercostal muscles. It takes air into and out of the lungs so that oxygen from the air can pass into the bloodstream and carbon dioxide can pass from the bloodstream into the air.

Energy production-respiration

We need to breathe to inhale fresh oxygen-rich air into our lungs to supply oxygen to the blood and to exhale carbon dioxide that has entered the lungs from the blood. The oxygen is needed for energy production in our cells and tissues. Carbon dioxide is the waste product of this process.

glucose + oxygen ————————> energy + carbon dioxide + water

This process of energy production in our cells is called aerobic respiration.

Anatomy of the lungs

There are two lungs-the left lung is slightly smaller than the right lung to make space for the heart which sits between them. The lungs are supplied with air from the windpipe or trachea. This connects the lungs with the nose and mouth. The trachea divides into the left and right bronchus, which supply each respective lung. These airways are reinforced with cartilage to keep them open during breathing. The bronchi then branch out like a tree into smaller tubes called bronchioles. These spread throughout the lungs and end in blind ending sacs called alveoli. The arrangement is sometimes called the bronchial tree. Pressed against the outside of the alveoli are thousands of small capillaries. The surface area of the alveoli is enormous-if they were squashed flat they would cover a tennis court.

Figure 2.1 The bronchial tree

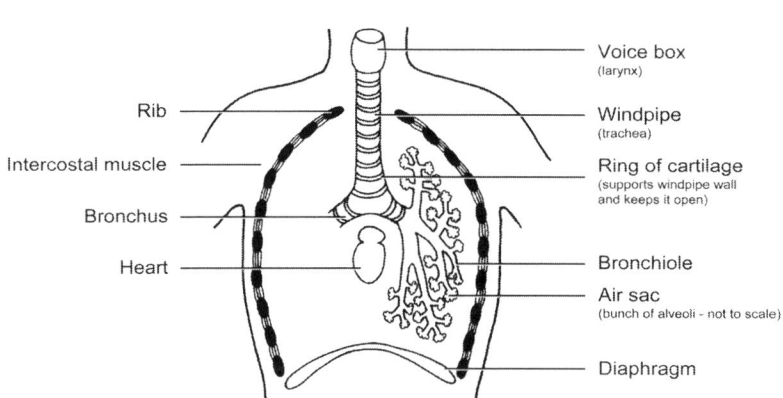

Gas exchange

The air enters the lungs via the tubes described above and enters the alveoli. These are lined with moisture to help gas exchange. There the oxygen in the air diffuses (a bit like percolation of coffee or tea) through the thin membrane of an alveolus and then through a thin capillary wall. From there it enters red blood cells down a concentration gradient and these are swept away by the circulatory system to the body's tissues. They supply the cells with oxygen for burning up glucose to provide energy. The waste gas from this process is carbon dioxide. It is carried in the blood plasma as hydrogen carbonate ions, back to the lungs. The carbon dioxide diffuses from the blood through the capillary walls in the lungs and into the alveoli. It is then breathed out.

Filtration of the air

The air breathed in through the nose is warmed and filtered by hairs. It is also saturated with water vapour. The trachea, bronchi and bronchioles are covered with ciliated cells with little hairs that remove dust and microbes from the breathing system. The dirt is trapped by mucus from mucus secreting cells and is swept upwards to the throat by the hairs (cilia), where it is swallowed.

Figure 2.2 Alveoli and bronchioles

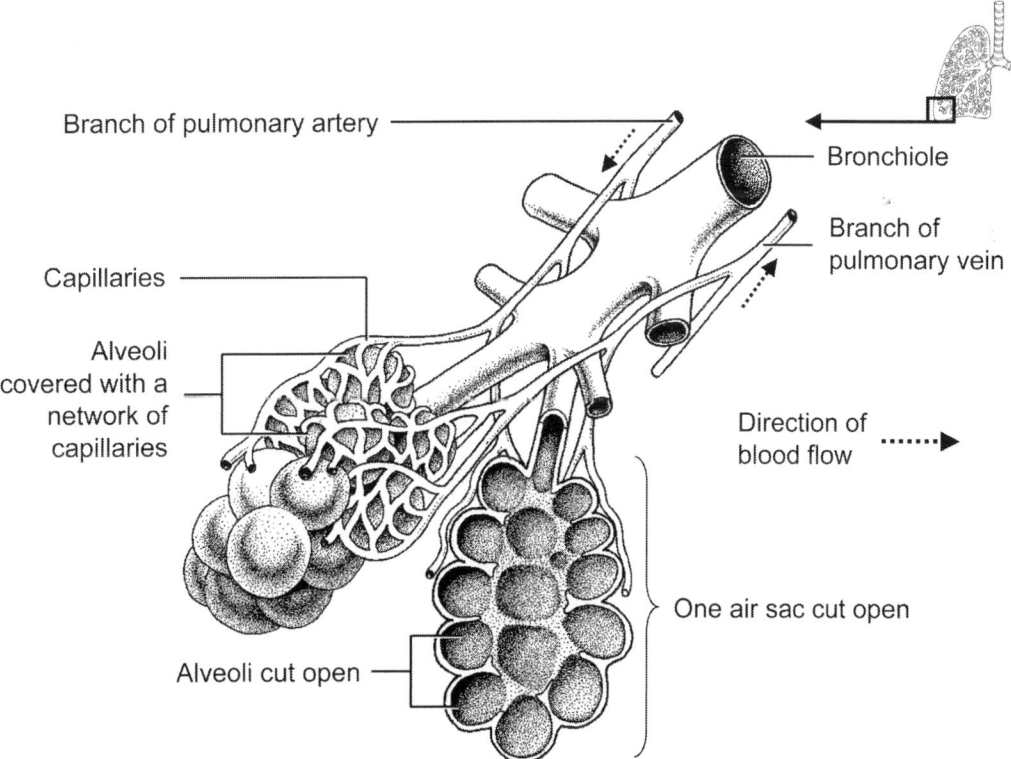

How we breathe

At the bottom of the thorax or chest is a large dome shaped muscle called the diaphragm. This separates the thorax from the abdomen. When we breathe in, signals from the brain cause the diaphragm muscle to contract. It flattens and moves downwards.

At the same time, under control of the brain, the intercostal muscles between the ribs contract and pull them upwards and outwards. The result of this is to expand the volume of the thorax. The pressure inside drops and air is sucked into the lungs– rather like a set of old fashioned bellows– causing inhalation.

Once the lungs have inflated enough, the stretch receptors in them send signals to the brain. These signal the end of inhalation. The diaphragm and intercostal muscles relax. The diaphragm springs back up and the ribs fall downward and inward. The volume of the thorax decreases and the pressure inside the lungs gets higher, squeezing the air out during exhalation.

Control of breathing

Carbon dioxide levels in the blood are detected by sensors, called chemoreceptors, in the carotid artery ands aorta. These stimulate the respiratory centres in the brain when carbon dioxide levels increase in the blood, e.g. as a result of exercise. The respiratory centres then speed up the rate and depth of breathing. This is how artificial resuscitation works– exhaled air is rich in carbon dioxide and is blown into the patients lungs to stimulate breathing. There is enough oxygen in the air to sustain life. (Also a large drop in oxygen levels in the blood will stimulate breathing.)

Exercise causes the rate and depth of breathing to increase and also the rate of the heart beat and volume of blood pumped out by the heart. These changes increase removal of carbon dioxide, lactic acid and heat from the muscle cells and increase oxygen and glucose supply to the muscles.

Anaerobic respiration

glucose ——————> lactic acid + energy

If vigorous exercise is performed, another energy system operates, known as anaerobic respiration. It produces small amounts of extra energy and does not use oxygen, but also produces lactic acid. This is toxic and causes muscle cramp, limiting the duration of exercise. After the exercise, faster and deeper breathing supplies extra oxygen to eliminate the lactic acid. This extra oxygen is called the oxygen debt. Fitter people produce less lactic acid and recover more quickly. This is the basis of fitness testing.

3 THE CIRCULATORY SYSTEM

The blood

Blood consists of a liquid medium, called plasma, containing dissolved chemicals and it carries in suspension solid cells. The plasma contains water. In this is dissolved glucose, proteins, hormones, fats, hydrogen carbonate, urea, vitamins and minerals.

The cells found in blood are erythrocytes or red blood cells, white blood cells or leucocytes and platelets.

The red blood cells are doughnut shaped and are full of an iron-rich pigment called haemoglobin. This carries oxygen around the body. When the red blood cells pick up oxygen from the lungs, this becomes oxyhaemoglobin. It releases the oxygen through the walls of the capillaries to the tissues and cells and reverts to haemoglobin. Red blood cells do not contain organelles or a nucleus.

There are two basic types of white blood cells. Their function is to fight disease. The lymphocytes are round and have a round shaped nucleus. They secrete antibodies which destroy viruses and bacteria. The other main category of white blood cell is the phagocyte. These are able to engulf bacteria and break them down.

Platelets look like fragments of red blood cells. They are essential for blood clotting.

Figure 3.1 Types of blood cells (drawn to scale)

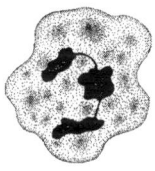

White blood cell
Neutrophil granulocyte leucocyte (granular cytoplasm and lobed nucleus) - also called phagocytes. The commonest type of white blood cell (60-70%). Kill bacteria in wounds, blood and tissue fluid (phagocytosis).

Red blood cell
Carry oxygen from lungs to all body tissues.
Face and side views of red blood cell.

Red blood cells as seen in a blood clot.

Platelets
Play a vital part in blood clotting.

The heart

This is a four chambered organ. It is essentially a double pump. The right side collects blood from the veins coming from the main organs and tissues and then pumps it to the lungs, where it is oxygenated. The left side receives blood from the veins coming from the lungs and pumps it to the arteries supplying the major organs and tissues. The receiving chambers of the heart are called atria. There is therefore a right and left atrium. The stronger pumping chambers are lower down and are called the right and left ventricles. The valves between the atria and ventricles are called atrio-ventricular valves (AV valves) and prevent the backflow of blood in the wrong direction. Sometimes the valve on the right side is called the tricuspid valve and that on the left side the bicuspid valve.

Control of the heart and the cardiac cycle

There is a region in the wall of the right atrium called the sino-atrial node. This gives off a regular electrical impulse that spreads across the atria causing them to contract. This phase of contraction is called atrial systole and pumps blood into the ventricles through the atrio-ventricular valves. The signal is delayed briefly at a node called the atrio-ventricular node, while the ventricles fill with blood. Next the signal travels down nerve fibres, called the Bundle of His, to the ventricles. These then contract as the signal spreads up them-this phase is called ventricular systole. The back pressure of blood closes the atrio-ventricular valves and the blood leaves the heart through the main arteries. To leave, the pressure has to be high enough to force open the semi-lunar valves. The heart then relaxes– this phase is called diastole. Hence, when blood pressure is measured there are two values– a high systolic pressure and a lower diastolic pressure. A typical reading is 120/80 mm mercury pressure.

Figure 3.2   Blood flow through the heart

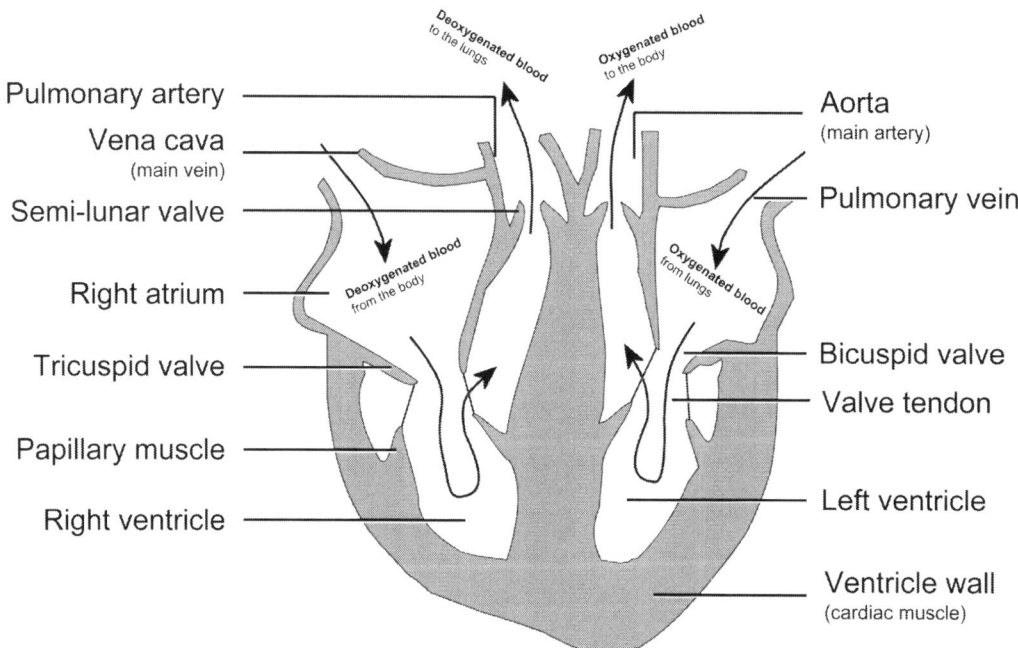

Figure 3.3 The cardiac cycle

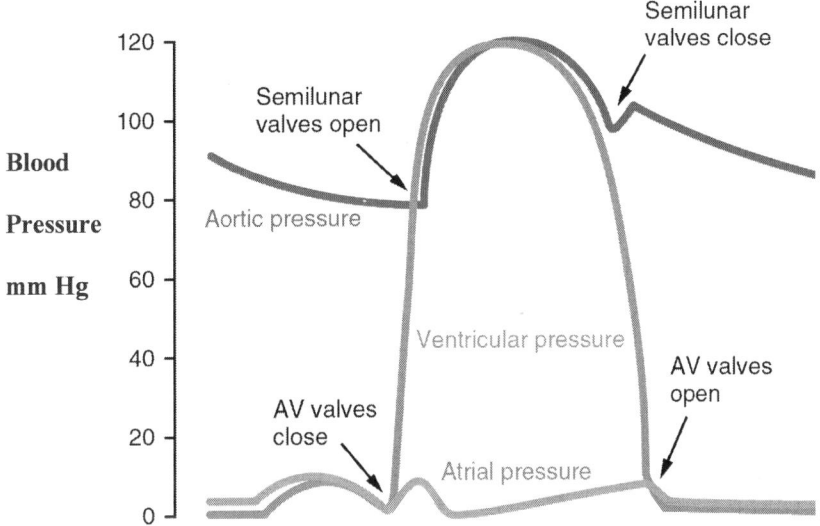

The electrocardiogram

The conduction mechanism in the heart can be used to diagnose heart problems. Leads are attached to the patient's chest, wrists and ankles. They are attached to a monitor and printer. The result is called an electrocardiogram or ECG.

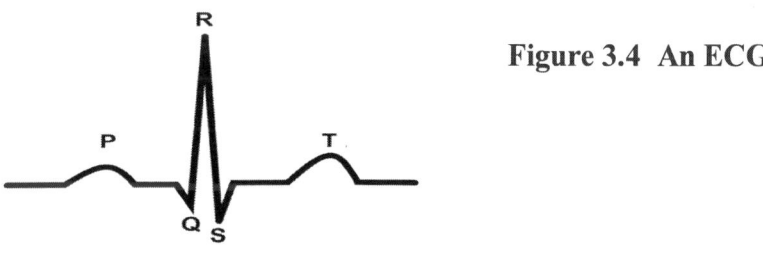

Figure 3.4 An ECG

The P wave corresponds to atrial contraction, the QRS wave to ventricular contraction and the T wave to relaxation of the ventricles. Any heart defects will alter the ECG– e.g. weak contractions of the atria or ventricles, or an irregular heartbeat (fibrillation).

Figure 3.5   The human circulatory system

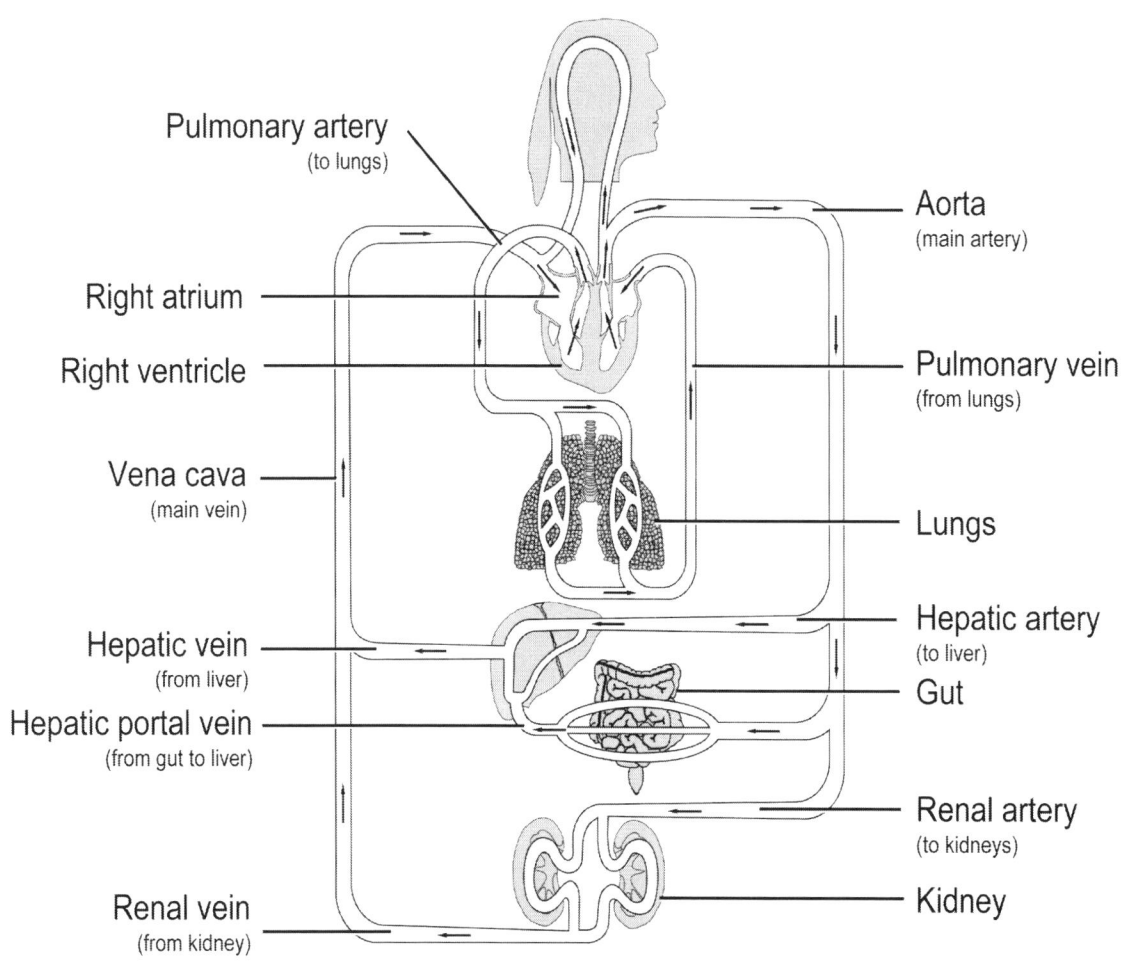

Blood Vessels

There are three main types of blood vessels- arteries, veins and capillaries. The arteries carry blood under high pressure from the ventricles of the heart to the major body organs and tissues. They have strong muscular walls, which ripple with the pulse when the heart contracts. They are able to withstand high pressures. (The arteries branch into smaller mini-arteries, called arterioles, which feed into the capillaries.)

The veins carry blood back to the heart from the organs and tissues. (They are supplied by mini-veins, called venules, which come from the capillaries.) Veins are wider than the arteries and since the blood pressure in them is low, have thinner walls. They contain valves to make sure that the blood flows in the correct direction, i.e. back to the heart. Skeletal muscles squeeze the veins during exercise, improving the blood flow back to the heart. Also, when we breathe in, the negative pressure in the chest helps draw blood back to the heart.

The main artery is the aorta, coming from the left ventricle. This feeds the renal artery to the kidneys, the hepatic artery to the liver and the main arteries to the other organs and the head, arms and legs. The pulmonary artery supplies deoxygenated blood to the lungs from the right ventricle. This blood is rich in carbon dioxide.

The main vein to the right atrium is the vena cava, coming from the veins from the organs, head, arms and legs. The pulmonary vein, carrying oxygenated blood, comes from the lungs to the left atrium.

The capillaries are tiny, thin-walled vessels found between the arteries and veins. They are present in huge numbers in the tissues and organs and are the site of exchange of nutrients between the blood and the cells. They also carry away waste products from the cells, such as carbon dioxide. Glucose, water and oxygen leave through the capillary walls and enter the cells. The fluid that leaves the capillaries from the blood is called tissue fluid and bathes the cells. It leaves the capillaries at the arterial end and is partly reabsorbed by them at the venous end and partly drained away by the lymphatic system (as lymph).

Figure 3.6 The structure of a capillary bed

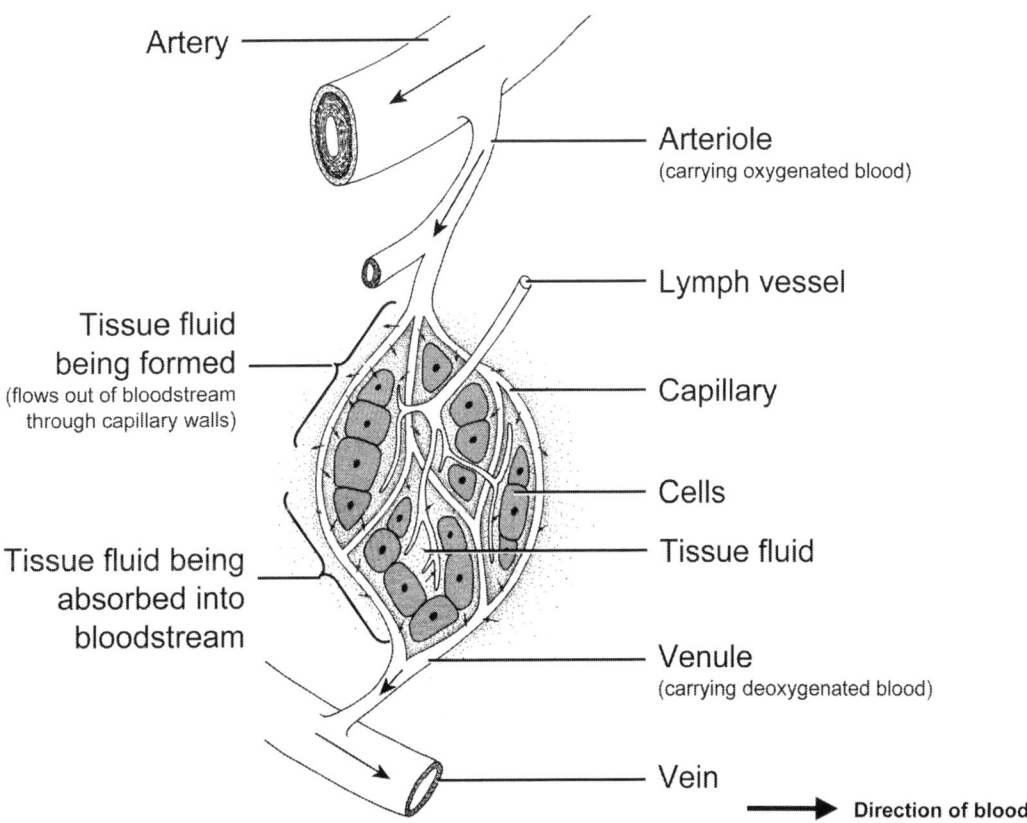

4 THE DIGESTIVE SYSTEM

The teeth

Large particles of food are broken down physically by the teeth. The sharp incisors and canine teeth at the front of the mouth bite and grasp the food, while the flatter premolars and molars grind the food up into a paste. This is called mechanical digestion.

Nutrition

There are five main groups of nutrients-carbohydrates, proteins, fats, vitamins and minerals. We also need water. Fibre is a type of carbohydrate made of cellulose and is indigestible. In this chapter we will look at the digestion of the first three types of nutrients.

Carbohydrates are large polysaccharides such as starch and cellulose. Starch is found in potatoes, flour and rice and is made of chains of glucose molecules. Glucose is a simple sugar called a monosaccharide. Another simple sugar is fructose, found in fruit and honey. Maltose, or malt sugar, is made of two glucose molecules and sucrose, or table sugar, is made of a glucose and a fructose molecule. These double sugars are called disaccharides.

Proteins are made of building blocks called amino acids—so called because they contain nitrogen, unlike carbohydrates. They are found in meat, fish and pulses.

Fats, or lipids as they are known, are made from molecules of fatty acids and glycerol. They are found in butter, lard and vegetable oils.

Digestion

The purpose of digestion is to chemically break down large, insoluble food molecules into simple, soluble molecules that are easily absorbed into the blood. The chemical breakdown is achieved by chemicals called enzymes.

Eating food is known as ingestion. In the mouth, the food is chewed into a paste and lubricated by saliva so that it is easily swallowed. At the same time, a carbohydrase enzyme in saliva (amylase) breaks starch into a sugar (maltose). The tongue forms a ball or bolus of food that is swallowed down the throat. The epiglottis is a flap of tissue which covers the windpipe so that the food is not breathed in accidentally.

The food passes down the gullet or oesophagus into the stomach. Involuntary action by longitudinal and circular muscles of the oesophagus squeeze the food down by a process called peristalsis– a bit like squeezing toothpaste out of a tube.

Inside the stomach, a protease enzyme (pepsin) is released from the gastric pits. Hydrochloric acid is also secreted from the stomach wall. The protease starts to break down proteins in the food into sections called polypeptides. The stomach also churns the food up into a soupy mixture called chyme.

After several hours, the food is released from the stomach and enters the small intestine. Here bile from the gall bladder is added to emulsify the fat into droplets with a large surface area. Just below the stomach is a feather-shaped organ called the pancreas. The pancreas releases the enzyme lipase into the small intestine to break fats into fatty acids and glycerol. The pancreas also releases more carbohydrase (amylase), which breaks starch down into sugars (maltose). Also it releases more protease (trypsin) to break proteins into polypeptides. The pancreas also releases an alkaline fluid to neutralise the acidic stomach contents, as the pancreatic enzymes work best in alkaline conditions.

Figure 4.1 The human digestive system

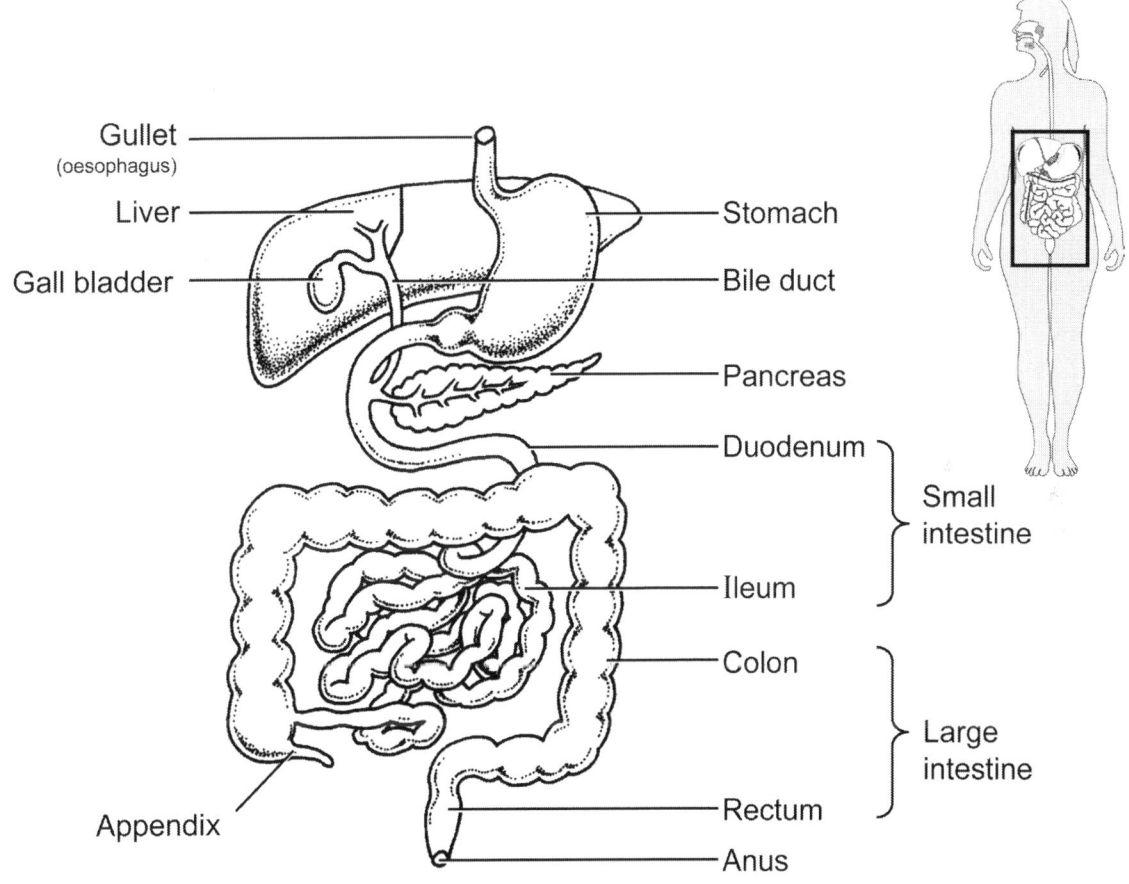

Absorption

The first part of the small intestine is called the duodenum. The second part is called the ileum. The latter has folded walls, with finger like projections called villi. The cells lining the villi have a brush border or microvilli to increase their surface area.

Glucose is absorbed firstly in the stomach. The cells lining the small intestine secrete enzymes that complete the chemical breakdown of digestion. Maltase is secreted to break the sugar maltose into glucose; sucrase breaks down sucrose and lactase breaks down lactose. The glucose is then absorbed across the lining of the ileum. Enzymes also break down the sections of previously partly digested proteins (polypeptides) into amino acids, which are also absorbed. The fatty acids and glycerol are also absorbed. The blood capillaries in the villi drain away the amino acids in the bloodstream. Absorption is a mixture of diffusion and a process involving energy called active transport.

Figure 4.2 A villus in the wall of the ileum

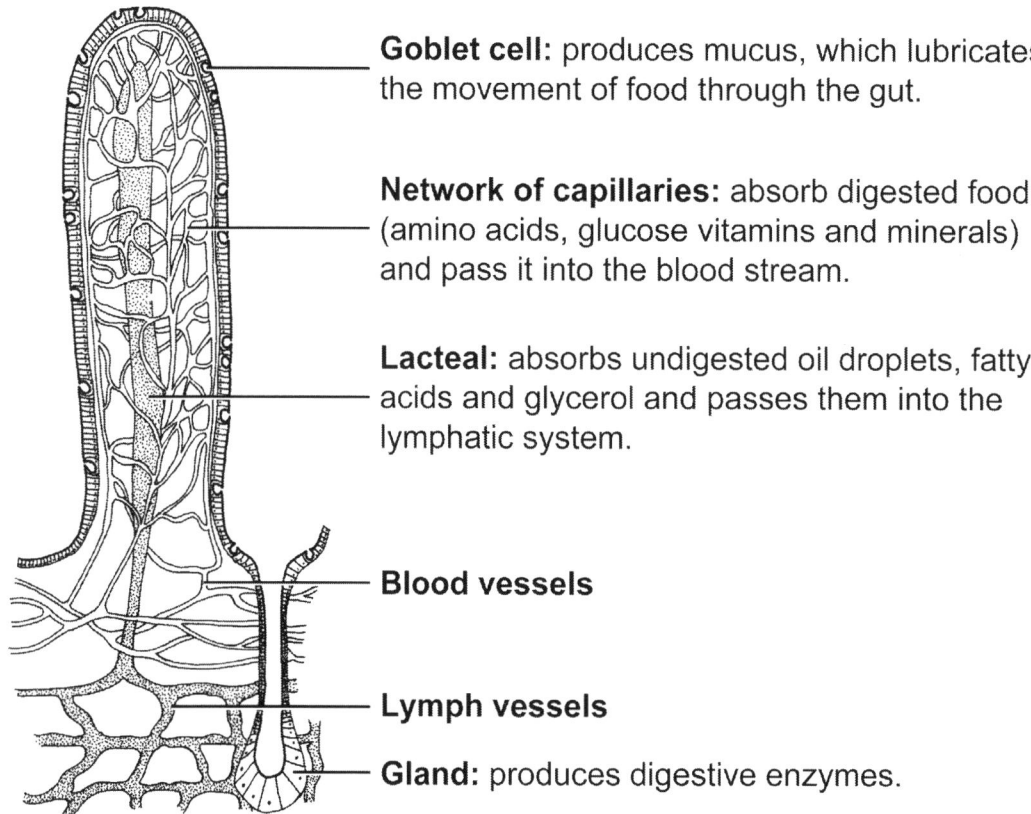

Assimilation of food products

The glucose and amino acids that have been absorbed from the gut are carried by the hepatic portal vein to the liver. The glucose is stored as a carbohydrate, called glycogen, in the liver cells or used for energy production. The amino acids are broken down into carbohydrate for energy and a waste product called urea, or used to build new amino acids to be used for protein synthesis in the body.

The fats are resynthesized in the cells of the villi from the absorbed fatty acids and glycerol and are drained away by the lymphatic system, which eventually returns them to the blood. The liver processes them too.

The large intestine

This is large in diameter, but short in length. Its main function is to absorb water from the intestine contents. The undigested food is then moved to the rectum where it is stored as faeces. Finally the faeces are expelled or egested through the anus.

SECTION TWO

DEFENCE AGAINST DISEASE

5 DISEASES

There are two categories of diseases– communicable diseases and non–communicable diseases. The former are spread from one person to another by harmful micro-organisms called pathogens. The latter are not passed from one person to another.

Communicable diseases

Bacteria

Diseases caused by bacteria include pneumonia, typhoid, diphtheria and tuberculosis. Bacteria are single-celled organisms which are capable of reproducing on their own. They can live in the bloodstream and body tissues.

Viruses

Diseases caused by viruses include the common cold, influenza, poliomyelitis, chicken pox, AIDS (acquired immunodeficiency syndrome), mumps, measles and rubella. Viruses are very small structures, which can only reproduce inside living cells. They make copies of themselves using the cell`s machinery and infect other cells.

HIV/AIDS

The human immunodeficiency virus (HIV) is transmitted in blood and during sexual intercourse in body fluids. It is called a retro-virus and damages the immune system via the white blood cells. It lies dormant and then causes full blown AIDS. It leads to death by opportunistic diseases which the body could normally destroy. It is treated with a drug called AZT.

Fungal diseases

These are spread by spores from one place to another, e.g. athletes foot, ringworm, thrush.

Parasites

e.g. malaria. This is caused by a single celled protozoan called plasmodium. It is spread by the female Anopheles mosquito and is transmitted via bites. It is injected by the mosquito via its saliva into the human bloodstream. It is common in tropical areas where the mosquito lives in water.

Non-communicable diseases

Dietary diseases

These are caused by poor diet or by malnutrition. Diseases caused by lack of vitamins are called deficiency diseases. Vitamin C is found in fresh citrus fruits and a lack of vitamin C causes scurvy. Wounds fail to heal and typically the gums bleed and teeth are lost. Lack of vitamin D causes rickets. The leg bones become curved. It is found in fish oils, butter and margarine. The action of sunlight on the skin makes vitamin D. Lack of vitamin A causes night blindness. It is found in carrots, milk and liver. Other deficiency diseases include lack of the B vitamins. These are found in yeast extract, liver and vegetables.

Lack of minerals in the diet can also cause deficiency diseases. Lack of calcium causes osteoporosis or brittle bones. Lack of iron causes anaemia. Good sources of calcium are milk and milk products and a good source of iron is red meat.

Heart disease

Coronary heart disease is caused by elevated cholesterol levels in the blood. Plaques of fatty substances are deposited in the coronary arteries of the heart and restrict blood flow to the heart muscle. A blood clot can form. In a heart attack, the heart muscle is starved of blood and nutrients and dies. The blockage can be relieved by inserting a metal scaffold called a stent to open the artery. The blood clot can be dissolved by drugs. These conditions are linked to an unhealthy diet, with too much saturated fat. High blood pressure can also damage the heart. Other heart conditions include inflammatory heart disease and valvular heart disease.

Smoking related diseases

The most well known is lung cancer. This is caused by the tar in tobacco smoke. Other diseases include heart disease and circulatory disorders, chronic obstructive pulmonary disease, and aggravated bronchitis. Tobacco smoke damages the alveoli and causes breathlessness or emphysema. Smoking tobacco can also cause sexual problems.

Cancer

This is caused by genetic mutations in certain genes called oncogenes, in the DNA of the cell. These changes make cells divide uncontrollably and these cells can then spread into other tissues and organs. This is called metastasis and produces secondary tumours. UV rays in sunlight can cause skin cancer and carcinogens in tobacco smoke cause lung cancer. Harmful chemicals and radiation cause cancer.

Psychiatric diseases

These can be broadly classified into two groups– neuroses and psychoses. Neuroses include obsessive compulsive disorder, phobias and anxiety. Psychoses are usually more serious and include bipolar disorder and schizophrenia (possibly linked to chemical imbalances in the brain).

6 IMMUNOLOGY: THE NON-SPECIFIC RESPONSE

The first line of defence

The skin

This has a dead, cornified outer layer of cells and prevents the entry of pathogens into the tissues. It is a barrier against infection, unless penetrated by insect bites or animal bites, e.g. malaria transmission from mosquitoes and rabies from dogs.

Stomach acid

The stomach contains hydrochloric acid at a pH of 2. It kills bacteria in the food and prevents food poisoning.

The respiratory system.

Hairs in the nose filter out large particles in the air breathed in. Also, goblet cells in the bronchial tubes secrete a sticky mucus that traps pathogens. Tiny hairs on the ciliated epithelium lining the bronchial tubes waft the mucus upwards to the throat, where it is swallowed.

The second line of defence

Blood clotting

If the skin is broken and a blood vessel ruptured, then there is a possibility that pathogenic bacteria or viruses may enter the wound. Therefore it is necessary for the blood to clot. This seals the wound and also prevents blood loss. If a blood vessel is damaged, platelets gather at the wound. These adhere to the cells and are said to be activated. They secrete thromboplastin, an enzyme that triggers blood clotting. This sets in motion a chain of reactions which end with the soluble protein, fibrinogen, carried by the blood, becoming solid strands of fibrin. These strands seal the wound. Red blood cells are trapped in the mesh of fibrin and form a blood clot. As the damaged skin heals, the clot shrinks pulling the wound together. Eventually the skin grows back and the blood clot is dissolved.

The inflammatory response

In the case of tissue damage or a localised infection, this response occurs. White blood cells, called mast cells, accumulate in the local area. These cells and ruptured cells release histamine. This causes vasodilation, increasing blood flow to the area and making it appear reddened. The permeability of the capillaries is increased and swelling due to increased tissue fluid formation occurs. The main benefit of this response is to increase the number of white blood cells travelling to the wound or infected area. These include phagocytes which ingest bacteria.

7 IMMUNOLOGY: THE SPECIFIC RESPONSE

The third line of defence

Types of specific immunity

There are two types– humoral, involving antibodies in the blood and cell-mediated.

Mechanism of humoral immunity

If bacteria enter the blood, marker proteins on their surface, called antigens, are presented on the outside of white blood cells called macrophages. These ingest the bacteria and display the antigens. This is called antigen presentation. Inside the lymph nodes, a white blood cell called a B-lymphocyte is selected according to the type of antigen presented. This is known as clonal selection. The B-lymphocyte then multiplies. This is called clonal expansion and causes large numbers of cells called plasma cells to be produced. These secrete chemicals called antibodies into the blood. Antibodies latch onto the surface antigens of all the bacteria of the type that invaded the blood and destroy them. There are millions of different types of B-lymphocytes in the body, each of which can produce a response to a different antigen and produce the necessary antibodies to lock onto the antigens.

How antibodies work

They work in different ways. Some clump the pathogens together– called agglutination. Others are anti-toxins which neutralise the poisonous toxins released by the bacteria. Some antibodies cause the bacterial cell walls to rupture, killing the bacteria– called lysis. All the products are then ingested by phagocytes.

Vaccination

An infection caused by a pathogen causes a primary immune response, generating antibodies from B- cells. Memory cells remember the type of infection, by remembering the antigens on the pathogen. They are ready to mount a huge secondary response if the pathogen re-enters the body. This is the principle of vaccination. A harmless bacteria or virus with the same antigens as the pathogen is used to stimulate an immune response and produce memory cells. If a deadly form of the disease vaccinated for enters the body, the immune system is primed by the memory cells and destroys the infection with the secondary response (huge numbers of antibodies are quickly generated)

Active immunity

Active immunity is stimulated by a vaccine or exposure to the pathogen and involves the body making its own antibodies against an antigen. It is usually long lasting.

Passive immunity

Passive immunity is when a person is given the antibodies, ready made, e.g. inside a pregnant mother antibodies are passed from the mother to the foetus. When a baby is breast fed, the mothers milk, in the first few days after child birth, contains large numbers of antibodies. The effect is short lived as the antibodies gradually get broken down in the bloodstream.

The cell-mediated response

This time another type of lymphocyte is involved, called a T-lymphocyte. These identify infected cells that the antibodies generated by the humoral response cannot penetrate to reach the pathogen. There are cells called T-killer cells that destroy infected cells, e.g. those infected by a virus. Also this system destroys cancer cells.

HIV and AIDS

If the immune system is damaged, e.g. by the HIV virus, the T- cells are destroyed and the victim is susceptible to pathogens normally fought off. These are called opportunistic diseases and can cause death. Also cancer cells are not destroyed and a common feature of AIDS sufferers is the presence of an otherwise rare skin cancer, called Kaposi`s sarcoma.

Organ transplants

Non-self antigens on foreign tissue introduced into the body stimulate an immune response. Again the cell-mediated system is involved. This causes transplanted organs to be rejected.

Tissue typing is done to ensure a good match of self and non-self antigens on the tissue. The ideal match would be from an identical twin, but this is rarely possible.

To prevent rejection of a transplanted organ, immune-suppressant drugs are used to reduce the immune response to the foreign tissue. The amount needed of these drugs depends on the match of antigens. A good match between donor and host reduces the amount of immune suppression needed. Obviously a poor match needs more immune suppression and the patient is then open to more infections.

SECTION THREE

HOMEOSTASIS

8 THE PRINCIPLES OF HOMEOSTASIS

The importance of homeostasis

Homeostasis is the maintenance of constant internal conditions within the body. Strictly speaking, they are kept within a limited range. In this book we will consider in detail temperature regulation and glucose control. However, other factors that are regulated include osmotic potential of the body fluids, ion control and blood pH.

Homeostasis is important, as the cells in the body need a relatively constant internal environment. The process resists external changes and allows the body to function in changing external conditions. Biochemical reactions are controlled by enzymes and changes in temperature and pH could damage or denature the enzymes, upsetting bodily functioning. Also, by keeping a constant water potential in fluid surrounding the cells, osmotic effects which could disrupt the cells are prevented.

How homeostasis works

Figure 8.1 Negative feedback

Stimulus ⟶ Receptor ⟶ Effector ⟶ System returned to the norm

Homeostasis involves receptors and effectors. Firstly, there are receptors which detect a deviation of the factor from the normal amount or norm and, secondly, a corrective mechanism involving effectors to bring the system back to the norm, e.g. in temperature regulation, temperature receptors measure the blood and skin temperature and effectors involving the skin bring the body temperature back to the normal value of 37 degrees Celsius.

The above process uses a mechanism called negative feedback. The changes from the norm are detected and measured and the system returned to the norm. Both increases and decreases from the norm can be detected.

9 TEMPERATURE REGULATION

Homeostatic mechanisms maintain a body temperature of 36.1 to 37.8 degrees Celsius inside the human body. Maintaining this temperature ensures that the biochemical reactions proceed at the correct rate and that the enzymes are not denatured.

The role of the hypothalamus and skin

The main control of body temperature is regulated and co-ordinated by the region in the brain known as the hypothalamus, which contains a thermostat.

Nerves from temperature receptors in the skin are connected to the brain and inform it of the external temperature, i.e. that of the surroundings. This information is fed into the voluntary nervous system and allows us to turn up the heat in a room, or move into warmer or colder surroundings.

The main method of temperature regulation is controlled by nerves going from the hypothalamus to the skin. There are two regions in the hypothalamus which are important in temperature regulation. They are supplied with blood at the body's core temperature. The first region is the heat conservation centre and controls warming mechanisms. The second region is the heat loss centre and controls cooling mechanisms.

Heat is lost from the skin by radiation, convection into the air or conduction into solid objects touching the body.

Figure 9.1 The structure of human skin

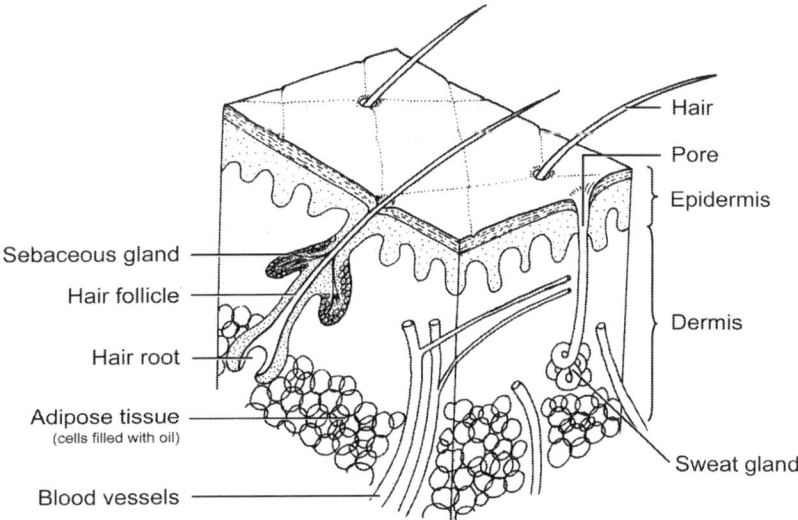

The response to cold

1. Shivering

The muscles in the body contract and relax alternately, generating heat as they do so. This helps warm the blood up.

2. Vasoconstriction

The blood supply to the skin is restricted. The arterioles supplying the skin capillaries constrict or narrow. This reduces the heat lost from the blood flowing to the skin. The skin becomes paler. Also shunt vessels open up, by-passing the skin capillaries and again reducing blood flow in the skin.

3. Hair Raising

This is of little value in humans and is more important in hairy mammals with thick coats of fur. The erector pili muscles contract, pulling the hairs upwards and trapping a layer of insulating air around the body. In humans it is seen as goose-pimples.

4. Increased Metabolic Rate

The hormone adrenaline is secreted by the adrenal glands. This increases the metabolic rate i.e. the body's rate of burning fuel, and generates heat. Over longer term exposure to cold, e.g. in cold climates, more of the hormone thyroxine is secreted by the thyroid gland, which also raises the metabolic rate.

The response to heat

1. Vasodilation

The arterioles supplying the skin`s capillaries open up or dilate, increasing the blood flow in the skin. Also the shunt vessels close down, increasing the blood flow to the capillaries in the skin. The increased blood flow to the skin loses heat by convection and radiation into the surroundings.

2. Sweating

This is a very effective way of losing heat, as 2,500 Joules of energy are required to evaporate one gram of sweat. Sweat is released from the sweat glands in the skin. It is mainly water and evaporates cooling the skin down.

3. Decreased Metabolic rate

The metabolic rate is reduced when we are hot as the body needs to generate less heat,.

Figure 9.2 Summary of temperature regulation in humans

Blood in hypothalamus too cold- Heat conservation centre activated- Responses:

Shivering

Vasoconstriction

Hair raising

Increased metabolic rate

Blood in hypothalamus too hot– Heat loss centre activated-Responses:

Vasodilation

Sweating

Decreased metabolic rate

Skin receptors provide additional information to voluntary area of brain.

Positive feedback and disorders

Hypothermia

If the body temperature falls too much and becomes dangerously low, e.g. by walking in the prolonged cold, then the usual homeostatic mechanisms fail and positive feedback occurs. As the temperature falls, the metabolic rate decreases. This then causes the body to become even colder. The system spirals downwards out of control below 27 degrees Celsius and, because the first organ to be affected is the brain, the subject may not even be aware of it to do anything about it, such as putting on warmer clothes. When the lower lethal temperature is reached death ensues.

Hyperthermia

If the body temperature rises too much, e.g. by walking in the hot sun, then again homeostatic control is lost and positive feedback occurs. The metabolic rate rises, increasing the core temperature even further. The system again spirals, this time upwards, leading to sunstroke and above 42 degrees Celsius death.

The above two situations are examples of positive feedback.

10 GLUCOSE REGULATION

Glucose is a sugar that is absorbed into the blood from the digestive system after consuming a meal. The blood travels to the pancreas. The norm for blood glucose is 80-100mg per 100ml blood.

The pancreas

The pancreas is a feather-shaped organ sited just below the stomach in the abdomen. Firstly, it is an exocrine organ and makes digestive juices released down the pancreatic duct into the small intestine. Secondly, it is an endocrine organ and makes hormones involved in the regulation of the concentration of glucose in the blood. It makes its hormones in the islet cells. These are called the Islets of Langerhans after the scientist who first discovered them. There are two types of islet cells– the alpha cells and the beta cells. The alpha cells secrete a hormone called glucagon into the blood and the beta cells secrete a hormone called insulin into the blood. Hormones are chemical messengers secreted by endocrine glands into the bloodstream.

If the blood glucose levels fall below the norm

This occurs as a result of fasting or exercise. The pancreas senses the drop in blood glucose levels and the alpha cells release glucagon into the blood. This is carried in solution in the blood plasma to the liver and exerts its main effect there. The enzymes in the liver are stimulated to break down its stores of the polysaccharide called glycogen into glucose. This is called glycogenolysis. This glucose is secreted into the blood restoring the glucose levels back to the norm. Also, in the body, other chemicals are converted into glucose in a process called gluconeogenesis.

If the blood glucose levels rise above the norm

This occurs after consuming a carbohydrate-rich meal. The pancreas senses the elevated levels of glucose flowing through the blood supplying it and the beta cells release the hormone insulin. This travels in the blood plasma in solution to the liver in the bloodstream and exerts its main effect there. The cells in the liver become more permeable to glucose and they absorb glucose from the blood. The enzymes in the liver cells are activated and the glucose is converted to the storage polysaccharide glycogen. This is termed glycogenesis. Thus, the blood sugar levels drop back to the norm. Also, other cells in the body are affected by insulin and they too absorb more glucose from the blood.

Problems with glucose control-Diabetes

There are two types of diabetes. Both cause glucose levels in the blood to be elevated. The first type is called Type One diabetes or Juvenile Onset diabetes and the second type is called Type Two diabetes or Maturity Onset diabetes.

Type One Diabetes

If the pancreas is damaged by an auto-immune reaction, usually in childhood, it can no longer secrete insulin. The blood levels of glucose become dangerously high. This causes glucose to be excreted in the urine and large volumes of dilute urine are produced by osmosis in the kidney tubules. The patient feels excessively thirsty and craves sweet foods. Often they are very tired.

The treatment is to inject insulin to replace that which is normally supplied by the pancreas. It is important for the patient to monitor their carbohydrate intake and to test their blood glucose levels every day. If they exercise, they will need less insulin as the muscles respire blood glucose causing the plasma levels to fall. Extreme rises or falls in blood glucose concentration can cause the person to enter into a coma.

Type Two Diabetes

This occurs later in life, usually above the age of 50. People predisposed to it are generally overweight and do little exercise. The pancreas supplies insulin, but the body cells are resistant to it. The blood sugar levels rise above the norm. It is treated firstly by dietary restriction of carbohydrates and then also by drugs, e.g. metformin hydrochloride.

Testing for Diabetes

This is done in hospital with the glucose tolerance test. The patient fasts overnight and then drinks a large flask of strong glucose solution in water. They have a blood sample taken every 30 minutes and a graph of blood plasma levels of glucose plotted. If the levels become very high and do not fall back to the norm of 80-100 mg glucose per 100ml blood within three hours, then the patient is diagnosed as diabetic.

SECTION FOUR

GENES AND GENE TECHNOLOGY

11 DNA AND GENES

DNA

The instructions for inheritance are found in the nucleus of a cell. The instructions are called genes and comprise lengths of a nucleic acid called deoxyribonucleic acid or DNA. The structure of this was elucidated by Watson and Crick at Cambridge University. DNA is a spiral shaped molecule called a double helix. It consists of two backbones of alternating sugar and phosphate molecules. The inner struts consist of two nitrogenous bases. One sugar, phosphate and base on each side is called a nucleotide. The sugar in DNA is deoxyribose.

Figure 11.1 The structure of DNA

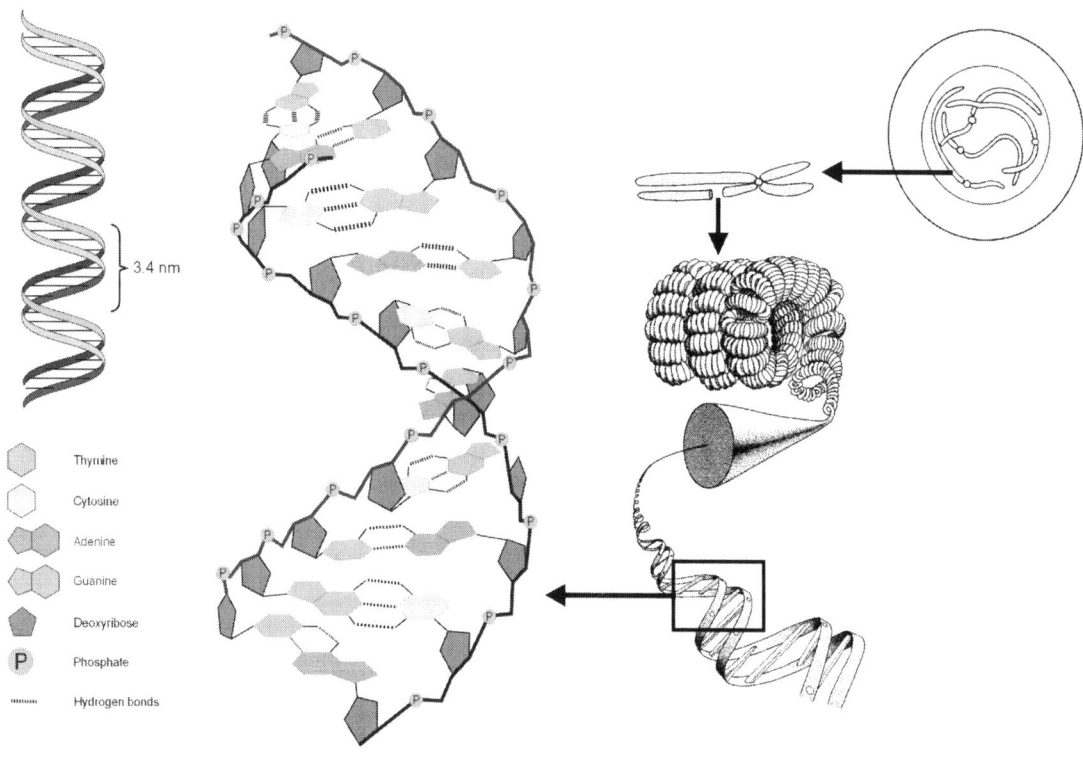

N.B. Ten base-pairs form one complete turn (360°) of the helix.

There are four possible bases– adenine (A), guanine (G), cytosine (C) and thymine (T). There are two bases at each section. Adenine always pairs with thymine and cytosine always pairs with guanine: the two bases are held by hydrogen bonds.

Protein synthesis

The whole DNA molecule is a giant genetic code. Every sequence of three bases on the one side, called the sense strand, codes for an amino acid. The triplet code is called a codon. The opposite strand is not used. The instructions comprising a gene are the genetic code for one protein.

The code is copied onto another nucleic acid called messenger RNA or ribonucleic acid. This process of copying is called transcription. RNA has a different sugar to DNA, namely ribose and also has the base uracil instead of thymine. It is single stranded. It forms a series of triplet base codes complementary to that in the DNA sense strand, e.g. A with T, C with G. The sense strand acts as a template for the messenger RNA.

The shorter messenger RNA molecule leaves the nucleus through the nuclear pores in the nuclear membrane and moves to a ribosome on the rough endoplasmic reticulum. There it controls the formation of a protein, using another type of nucleic acid, transfer RNA. This carries individual amino acids to the ribosome, where they are assembled into a protein. This process is called translation.

DNA Replication

Figure 11.2 DNA Replication

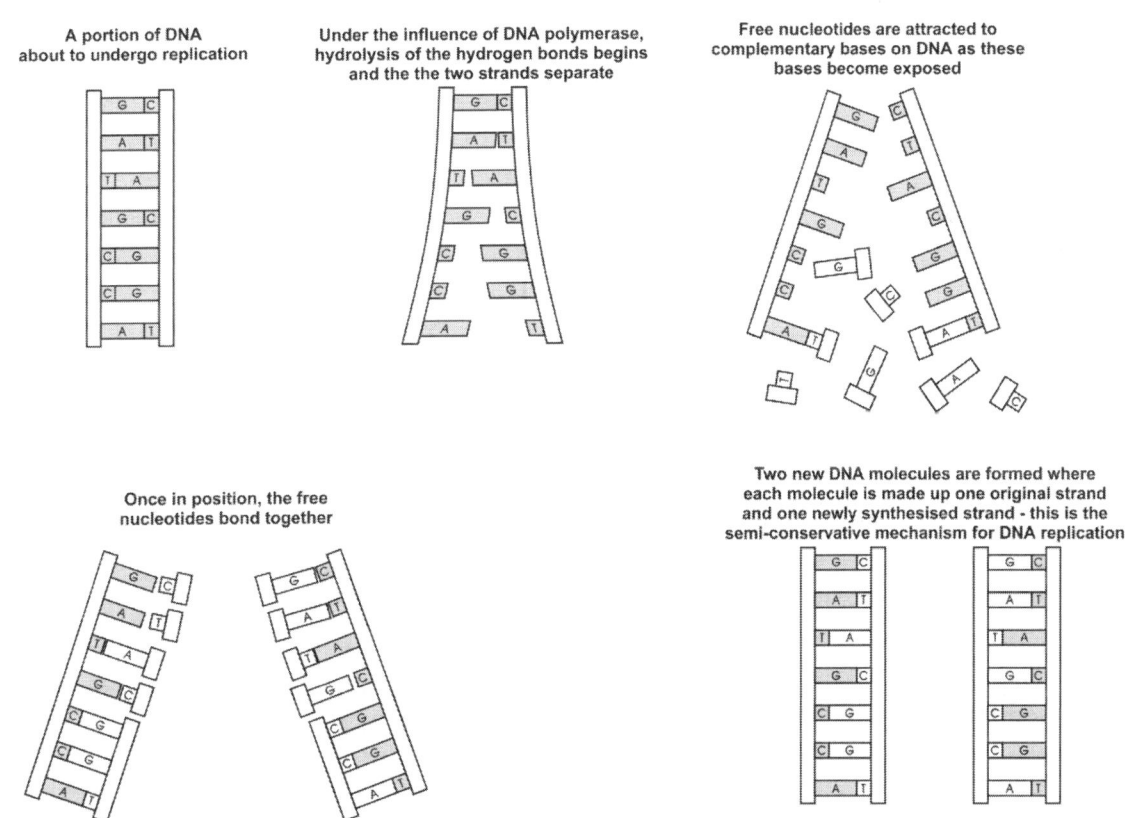

The DNA is dispersed throughout the nucleus during a stage called interphase. At the start of cell division, the DNA condenses into visible threads called chromosomes. Prior to cell division, during interphase, the DNA replicates and forms a copy of itself, so that when the chromosomes are visible they appear double.

During replication of the DNA, it unzips into two separate strands. The hydrogen bonds between the bases unlock and a new strand is formed alongside the original strand, under the action of the enzyme DNA polymerase. The free nucleotides join together and onto their complementary nucleotide base sequence. This is called semi-conservative replication, because we end up with one old strand of nucleotides and one new strand of nucleotides in each DNA molecule.

12 CELL DIVISION

There are two types of cell division– mitosis and meiosis.

Mitosis

During growth of the body and also during replacement and repair of tissues in the body, cells are constantly dividing. The process of cell division involved is called mitosis. There are several stages to this process. The daughter cells produced are identical to the parent cell.

In interphase, the DNA is spread throughout the nucleus. Just prior to cell division it replicates. During the first stage, called prophase, the DNA condenses into threads called chromosomes. In the first stage, the chromosomes appear double and there are two threads, or chromatids, visible for each chromosome. The twin chromatids line up on the equator of the cell in the next stage, called metaphase, attached to spindle fibres. These then pull the twin chromatids apart during anaphase and, finally, the separated chromatids move to opposite poles of the cell during telophase. Each chromatid is now called a chromosome.

The spindle fibres disappear and the nuclear membranes appear around the new sets of chromosomes. Then the cell membrane and cytoplasm divide and two new cells are formed. They have exactly the same number of chromosomes as the parent cell, i.e. 46. Then the chromosomes disperse and the DNA is spread again throughout the nucleus. This is the end of mitosis.

Figure 12.1 Mitosis

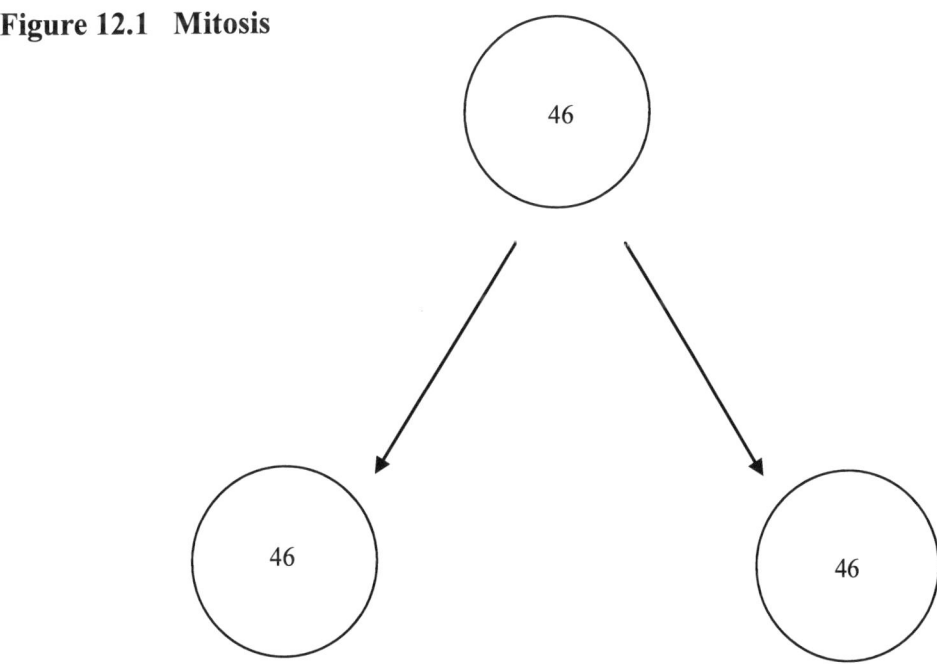

Meiosis

This is the type of cell division that occurs to produce the gametes or sex cells. In the testes, the spermatozoa are produced and, in the ovaries, the ova or egg cells are produced. They each have only 23 single chromosomes in their nucleus compared to 46 in a normal body cell.

In meiosis, the DNA condenses into chromosomes as in mitosis, and also again, during interphase, the DNA replicates prior to cell division.

However, in meiosis there are two cell divisions, not one as in mitosis. In the first cell division, homologous pairs of chromosomes separate and, in the second stage, the chromatids separate.

Finally, the nuclear membranes reform and the gametes are produced. They each have their own combination of genes and are all different.

Figure 12.2   Meiosis

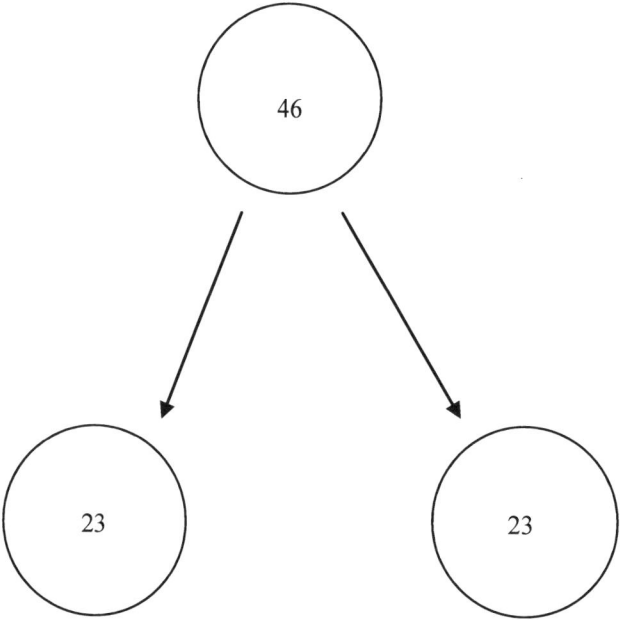

The life cycle of humans

Each sperm and egg cell only has 23 chromosomes in its nucleus. When, in sexual reproduction, the sperm fertilises an egg cell, the nuclei of the two gametes fuse and a new cell is formed called a zygote. This has 46 chromosomes in its nucleus.

In the human the number of 23 chromosomes is sometimes called n, or haploid, and 46 chromosomes is called 2n, or diploid.

Finally, the zygote divides billions of times over by mitosis to create a new human being.

Figure 12.3

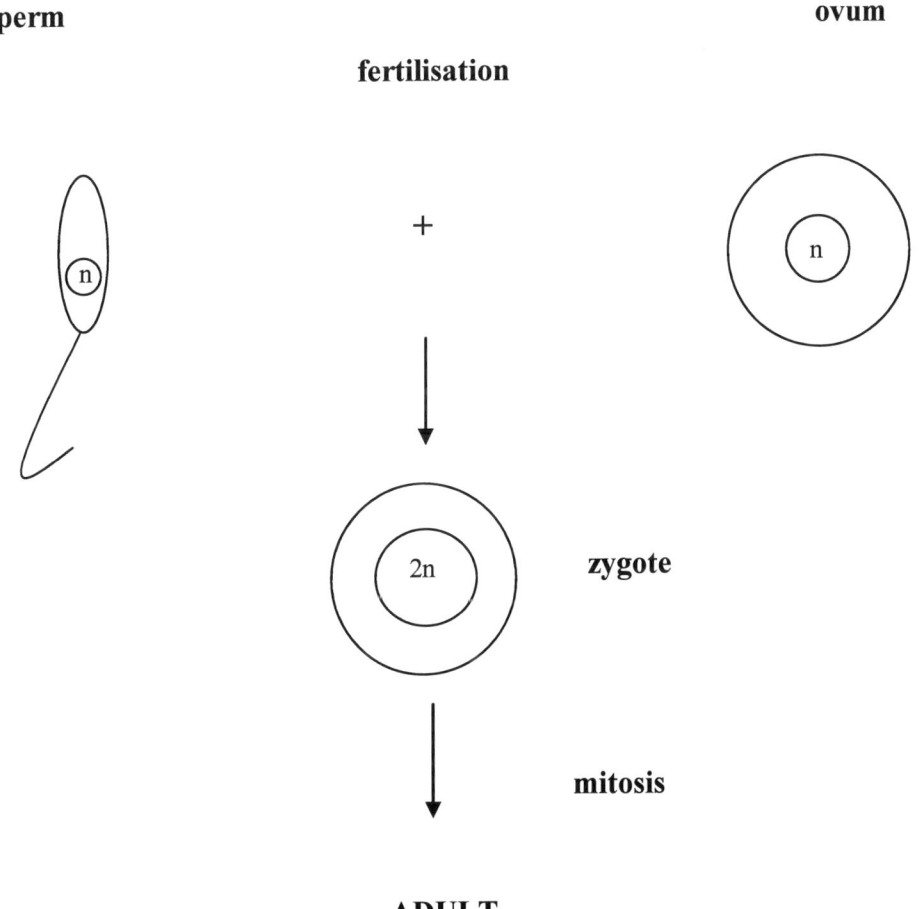

13 GENETICS

Mendel's pea plants-monohybrid crosses

In the nineteenth century, a monk called Mendel carried out the first genetics experiments on pea plants in his monastery garden. His experiments laid the foundation of modern genetics and are still relevant today.

He first of all bred true-breeding pea plants with a red flower. These were crossed with other true-breeding plants with a red flower and always produced pea plants with red flowers. He also cross bred true-breeding pea plants with white flowers with each other.

Mendel then cross bred the true-breeding red flowered pea plant with the true-breeding white flowered pea plant. The result was that all the offspring had red flowers. He realised that there were units of inheritance, which we now call genes, for the red and white colours. The red gene was dominant to the white gene. The experiment can be represented by a genetic diagram.

Figure 13.1   The monohybrid cross

R denotes the dominant allele (for red colour)

r denotes the recessive allele (for white colour)

P1 phenotypes	Red		White
P1 genotypes	RR	x	rr

Gametes	R	R	r	r
F1 genotypes	Rr	Rr	Rr	Rr

F1 phenotypes : All Red

We have represented the gene for red colour with a capital letter R and the gene for white colour with a lower case r. These pairs of genes are called alleles. The one gene came from the chromosome given by the male gamete and the other gene from the chromosome given by the female gamete. The pair of chromosomes they come from are called homologous chromosomes. The capital R represents the dominant gene and the little r the recessive gene. The first parental generation is called the P1 generation and the offspring the F1 generation. The genetic names are called genotypes. The physical expression of the genes, i.e. the colours, are called phenotypes.

Mendel then crossed the F1 generation.

P2 phenotypes   Red   Red

P2 genotypes   Rr   x   Rr

Figure 13.2

Gametes	R	r
R	RR	Rr
r	Rr	rr

F2 genotypes   RR : 2Rr : rr

F2 phenotypes   3:1 Red : White

For convenience, we have used a Punnett Square to carry out the cross. The new offspring of this second cross are called the F2 generation. We can see that the F2 generation has the following genotypes- RR, Rr and rr. The ratio of numbers of different coloured plants were 3 red to 1 white.

The RR genotype is called homozygous or double dominant, the rr genotype is called double recessive and the Rr genotype is called heterozygous.

Human genetics

We can now apply Mendel's principles to human genetics.

Huntington's disease

This is a dominant disorder and follows exactly the same pattern as Mendel's red and white flowered pea plants. The dominant gene is represented as H and the recessive gene by h. The disorder is caused by the dominant gene H and codes for the production of a certain protein. This causes mental and nervous degeneration and death after the age of 40 years. The condition is passed on through the generations as individuals do not realise they have the gene until after their child bearing years.

Sufferer   genotype- HH

Sufferer   genotype- Hh

Healthy person   genotype- hh

e.g. If a heterozygous sufferer has children to a healthy person, what are the chances of their children having the disease?

Answer-   **Hh x hh**

Figure 13.3

Gametes	H	h
h	Hh	hh
h	Hh	hh

Thus it can be seen that the children have a 50% chance of inheriting the disorder.

Cystic fibrosis

This is different to the red and white flowered pea plants as it is caused in humans by a recessive gene. Lack of the dominant gene means that the epithelial cells lining the lungs, reproductive tubes and pancreatic duct are unable to make the correct carrier protein in their cell membranes. This results in faulty chloride ion transport and a thick sticky mucus blocks the lungs and these tubes.

Healthy person   genotype-   FF

Carrier, but healthy genotype-   Ff

Sufferer   genotype-   ff

e.g. If two carriers have children, what would be the chance of them having the disorder?

Answer-

Ff x Ff

Figure 13.4

Gametes	F	f
F	FF	Ff
f	Ff	ff

i.e. 25% chance of having the disease and a 50% chance of being a carrier.

Sex Determination in humans

There are two sex chromosomes, X and Y. The Y chromosome is only found in males and is much shorter than the other chromosomes.

Females have the following genotype   XX

Males have the following genotype   XY

The sperm and eggs can only carry one of the two sex chromosomes. Sperm can carry either an X or a Y chromosome. The egg or ovum always carries an X chromosome. Hence, it can be seen that it is the sperm that decides the sex of the child produced by fertilisation of the ovum.

X denotes the female chromosome

Y denotes the male chromosome

Parental genotypes   XY   x   XX

Gametes   X   Y   X   X

Figure 13.5

Gametes	X	X
X	XX	XX
Y	XY	XY

Offspring genotypes   2XY : 2XX

Offspring phenotypes   2 male : 2 female   i.e. 50% chance of a boy or girl.

Sex-linkage

The short length of the Y chromosome makes it unable to carry as many genes as the X chromosome. There are some disorders, called sex-linked disorders, which are only carried on the X chromosome. They are caused by recessive genes on the X chromosome. Examples include haemophilia, in which an individual is unable to produce Factor V111 for normal blood clotting, and also some types of colour blindness.

Healthy male   genotype- $X^H Y$

Male sufferer   genotype- $X^h Y$

Healthy female   genotype- $X^H X^H$

Female carrier   genotype- $X^H X^h$

Female sufferer   genotype- $X^h X^h$

The genetic crosses are exactly the same as for sex determination. The genes are carried on the sex chromosomes and the above genotypes indicate whether the person is a carrier, sufferer or healthy.

Pedigrees or family trees can be produced to show the inheritance of a sex-linked disorder e.g. haemophilia. The following shows an example.

Figure 13.6 Inheritance-Haemophilia Pedigree

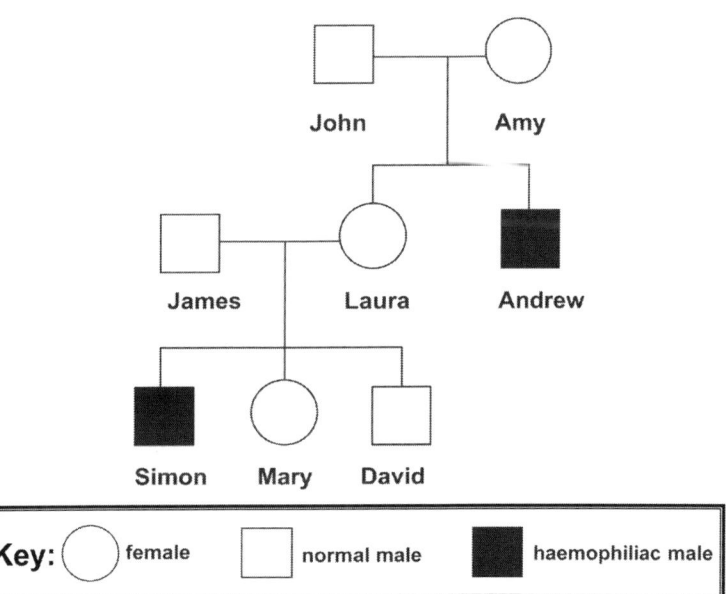

ABO Blood Groups

This is an example of multiple alleles. There are three possible genes for blood type. There are O, A and B genes. Only two can be carried by an individual at any one time and one is inherited from each parent. The genes A and B are dominant to O and the following blood groups are possible:

Blood group	**Genotype**
A	AA or AO
B	BB or BO
AB	AB
O	OO

Example of inheritance-

AA x BO

Figure 13.7

Gametes	A	A
B	AB	AB
O	AO	AO

ie 50% children are group AB and 50% are group A

Incomplete dominance

So far we have considered the cases of complete dominance, i.e. where one allele is completely dominant over a recessive gene. However, in sickle cell anaemia both alleles are expressed.

Normal allele- A

Sickle cell allele- a

Healthy person   genotype- AA

Carrier of sickle cell anaemia   genotype- Aa

Sufferer of sickle cell anaemia   genotype- aa

Inheritance is the same pattern as for the monohybrid cross with pea flowers.

Sickle cell anaemia causes the red blood cells to have a sickle shape instead of a round shape. They are less able to carry oxygen and tend to block up capillaries, leading to bone infections. The sickness is characterised by lack of breath and joint pain.

The double recessive sufferer has severe symptoms of the disease. The carrier, however, still has 50% of their red blood cells affected. This disorder is very common in Africa, because it gives protection against malaria and people carrying the recessive gene, a, do not die from this common tropical disease.

14 GENE TECHNOLOGY

Gene technology is more commonly known as genetic engineering. It is used to transform or change the DNA of micro-organisms to make them produce useful chemicals. It is commonly used to produce bacteria that will secrete human insulin or, with different genes, human growth hormone. There are other types of gene technology used to produce plants with new properties which give better yield, herbicide resistance, or the ability to grow in arid regions of the world. In this chapter we will look at how human insulin is produced. Also we will briefly look at gene therapy in medicine.

Producing human insulin

There are several stages to this process. We will look at them in turn.

Stage One

A suitable bacterium is taken, its cell wall broken open and the circular loop of DNA inside removed and saved. The loop of bacterial DNA is called a plasmid.

Stage Two

The human gene for insulin production is identified on a pancreatic chromosome using a gene probe– a series of complementary bases to the gene being located, attached to a marker.

Stage Three

The plasmid is cut open at a certain base sequence using a restriction endonuclease enzyme. The human gene is cut from the pancreatic chromosomal DNA using the same type of enzyme to ensure the same base sequences, known as sticky ends, are at the ends of the cut DNA.

Stage Four

The human gene is spliced into the plasmid using a ligase enzyme.

Stage Five

Often a gene for antibiotic resistance is also added in a similar fashion to the above to the plasmid.

Stage Six

The plasmid is put back into a bacterium. Often large numbers of transformed plasmids are mixed with bacteria. They are subjected to ice-cold calcium chloride solution and then warmed to make them take up the plasmids. Only a few bacteria will take up plasmids.

Stage Seven

The bacteria are grown on agar nutrient jelly impregnated with antibiotics. This will kill any bacteria that did not take up the spliced genes. The ones that took up the new plasmids will be resistant to antibiotics and will also contain the gene for human insulin production.

Stage Eight

The viable bacteria are cultured in sterile media and put into a large tank called a fermenter. Nutrient solution is added and the mixture gently stirred. The temperature is carefully regulated and oxygen supplied.

Stage Nine

The bacteria multiply asexually, making clones of themselves that are genetically identical. They each secrete human insulin by reading the new DNA added to them.

Stage Ten

The liquid is drained from the fermenter and the bacteria filtered out. The human insulin is dissolved in the liquid, which is then purified for medical use.

Human insulin prepared in this way is cheaper to manufacture than the previous method of extraction of insulin from the pancreas of pigs or cattle and has less side effects.

Gene therapy

Cystic fibrosis is caused by a faulty gene in epithelial cells. It is a recessive disorder. The gene that codes for the CFTR carrier protein mutates. (Cystic fibrosis transmembrane regulator protein). As a result there are too many chloride ions inside the cells. The result is a thick, sticky mucus that blocks the lungs, the reproductive tubes and the pancreatic duct. The main treatment is by physiotherapy. However, a new treatment involving gene therapy is being trialled. A cloned gene for normal CFTR is inserted into a virus. The virus is sprayed into the lungs and enters the epithelial cells where its DNA causes the cells to produce the healthy carrier protein. Another method being tried is to package the healthy CFTR gene in lipid droplets called liposomes. These are sprayed into the lungs of the sufferer and penetrate the epithelial cells. So far results have been disappointing.

INDEX

A

Absorption, 18
Active transport, 18
Adenine, 36
Adrenaline, 30
Aerobic respiration, 8
AIDS, 22
Allele, 36
Alpha cell, 32
Alveolus, 9
Amino acid, 16
Anaemia, 23
Anaerobic respiration, 10
Anaphase, 39
Antibody, 25
Antigen, 25
Anus, 17
Aorta, 13
Arteriole, 14
Artery, 14
Atrium, 13
AV valve, 13

B

Bacterium, 22
Base, 36
B-cell, 25
Beta cell, 32
Bicuspid valve, 12
Bile, 17
Bipolar disorder, 23
Blood groups, 48
Blood pressure, 12
Bolus, 16
Bronchiole, 8
Bronchus, 8
Bundle of His, 12

C

Cancer, 23
Capillary, 15
Carbohydrase, 16
Carbohydrate, 16
Cardiac cycle, 12
Cells, 6
Centriole, 6
Chemoreceptors, 10
Chicken pox, 22
Chromatid, 39
Cilia, 24
Colon, 19
Colour blindness, 47
Cystic fibrosis, 45, 51
Cytoplasm, 6
Cytosine, 36

D

Diabetes, 33
Diaphragm, 10
Diastole, 12
Diffusion, 18
Diploid, 41
Diphtheria, 22
Disaccharide, 16
Dominant, 42
DNA, 36
DNA polymerase, 37
Double helix, 36
Duodenum, 17

E

Ecg, 13
Effector, 28
Egestion, 19
Emphysema, 23
Endocrine gland, 32
Endoplasmic reticulum, 6
Energy, 8
Enzyme, 16
Epiglottis, 16
Exocrine gland, 32

F

Fatty acids, 16
Fertilisation, 41
Fibrin, 24
Fibrinogen, 24
Fitness testing, 10
Fructose, 16

G

Gall bladder, 17
Gametes, 41
Gene, 36
Genetics, 42
Glucagon, 32
Glucose tolerance test, 33
Gluconeogenesis, 32
Glycerol, 16
Glycogen, 32
Glycogenolysis, 32
Golgi body, 6
Guanine, 36

H

Haemoglobin, 11
Haemophilia, 47
Hair, 29
Haploid, 41
Heart, 12
Hepatic artery, 14
Heterozygous, 43
Histamine, 24
Homozygous, 43
Hormone, 32
Huntington`s disease, 44
Hypothalamus, 31
Hyperthermia, 31
Hypothermia, 31

I

Ileum, 17
Immune-suppression, 26
Influenza, 22
Insulin, 32
Intercostal muscles, 10
Interphase, 39
Ischaemic disease, 23
Islets of Langerhans, 32

K

Kaposi`s sarcoma, 26

L

Lacteal, 18
Lactic acid, 10
Large intestine, 19
Ligase, 50
Lipase, 17
Lipid, 16
Liver, 19,32
Lymphatic system, 15
Lymphocytes, 11
Lysosome, 6

M

Macrophage, 25
Malaria, 22
Maltose, 16
Maltase, 17
Meiosis, 40
Memory cells, 25
Metabolic rate, 30
Metaphase, 39
Metastasis, 23
Mitochondria, 6
Mitosis, 39
Monosaccharide, 16
Mouth, 16

N

Negative feedback, 28
Neuroses, 23
Norm, 28
Nucleotide, 36

O

Opportunistic diseases, 22, 26
Organs, 7
Organ systems, 7
Osteoporosis, 23
Oxygen debt, 10

P

Pancreas, 17
Passive immunity, 26
Pedigree, 47
Pepsin, 16
Peristalsis, 16
Phagocytes, 11
Plasma, 11
Plasmid, 50
Plasmodium, 22
Platelet, 11
Poliomyelitis, 22
Positive feedback, 31
Prophase, 39
Protease, 16
Protein, 16
Pulmonary artery/vein, 14
P-wave, 13

Q

QRS complex, 13

R

Receptor, 28
Recessive, 42
Rectum, 19
Renal artery/ vein, 14
Respiration, 8
Restriction enzyme, 50
Ribs, 10
Rickets, 23
Ribosome, 6
RNA, 37

S

Schizophrenia, 23
Semi-conservative replication, 37
Semi-lunar valve, 12
Shivering, 30
Shunt vessels, 30
Sickle cell anaemia, 49
Skin, 29
Small intestine, 17
Smoking related disease, 23
Stent, 23
Stomach, 16
Stretch receptors, 10
Sweating, 30
Systolic pressure, 12

T

T-cell, 25
Telophase, 39
Thorax, 8
Thromboplastin, 24
Thrush, 22
Thymine, 36
Thyroxine, 30
Tissues, 7
Tissue fluid, 15
Tricuspid valve, 12
Tuberculosis, 22
T-wave, 13
Typhoid, 22

U

Uracil, 37
UV light, 23

V

Vaccine, 25
Vasoconstriction, 30
Vasodilation, 30
Vein, 14
Vena Cava, 14
Ventricle, 12
Villus, 18
Virus, 22

W

White blood cells, 11

Z

Zygote, 41

Printed in Great Britain
by Amazon.co.uk, Ltd.,
Marston Gate.